CODE READER™
Making Difficult Words Easy

Code Reader Books provide codes with "sound keys" to help read difficult words. For example, a word that may be difficult to read is "unicorn," so it might be followed by a code like this: unicorn *(YOO-nih-korn)*. By providing codes with phonetic sound keys, Code Reader Books make reading easier and more enjoyable.

Examples of Code Reader™ Keys

Long a sound (as in make):
a *(with a silent e)* or **ay**
Examples: able *(AY-bul)*; break *(brake)*

Short i sound (as in sit): **i** or **ih**
Examples: myth *(mith)*; mission *(MIH-shun)*

Long i sound (as in by):
i *(with a silent e)* or **y**
Examples: might *(mite)*; bicycle *(BY-sih-kul)*

Keys for the long o sound (as in hope):
o *(with a silent e)* or **oh**
Examples: molten *(MOLE-ten)*; ocean *(OH-shen)*

Codes use dashes between syllables *(SIH-luh-buls)*, and stressed syllables have capital letters.

To see more Code Reader sound keys, see page 44.

DOLPHINS AND SEA TURTLES

Treasure Bay

Dolphins and Sea Turtles
A Code Reader™ Chapter Book
Blue Series

This book, along with images and text, is published under license from The Creative Company. Originally published as Spotlight on Nature: Dolphin and Spotlight on Nature: Sea Turtle © 2020 Creative Education, Creative Paperbacks

Additions and revisions to text in this licensed edition:
Copyright © 2025 Treasure Bay, Inc.
Additional images provided by iStock

All rights reserved.

Reading Consultant: Jennifer L. VanSlander, Ph.D., Asst. Professor of Educational Leadership, Columbus State University

Code Reader™ is a trademark of Treasure Bay, Inc.

Patent Pending.
Code Reader books are designed using an innovative system of methods to create and include phonetic codes to enhance the readability of text.
Reserved rights include any patent rights.

Published by
Treasure Bay, Inc.
PO Box 519
Roseville, CA 95661 USA

Printed in China

Library of Congress Control Number: 2024944836

ISBN: 978-1-60115-721-8

Visit us online at:
CodeReader.org

PR-1-25

CONTENTS

CHAPTER 1 Meet the Family 2

CHAPTER 2 Welcome to the World 8

CHAPTER 3 Mealtime 16

CHAPTER 4 Staying Safe 24

CHAPTER 5 Cool Facts 32

CHAPTER 6 Family Album Snapshots 38

Glossary 42

Questions to
Think About 43

Sounds Keys for Codes 44

CHAPTER ONE

MEET THE FAMILY

It's been said that there are plenty of fish in the sea. But did you know there are also lots of mammals *(MAM-mulz)* and reptiles?

Two favorites *(FAY-vrits)* are dolphins and sea turtles!

In addition to their outer eyelids, sea turtles have a see-through inner eyelid that can close over each eye. These see-through eyelids act like swimming goggles.

CLOSE-UP
Eyesight
Dolphins have excellent vision *(VIZH-un)*. Their eyes can move independently of each other, allowing the dolphin to look in two different directions at once *(wuns)*. Dolphins can also use both eyes together to focus *(FOH-kus)* and judge distance.

A sea turtle's body is protected by armor. The top covering is called the carapace *(KAIR-uh-pase)*. The bottom is called the plastron *(PLAS-trun)*. These pieces *(PEE-sez)* are held together by bones called bridges.

SEA TURTLE DIAGRAM

Carapace
(KAIR-uh-pase)

flipper

beak

Plastron
(PLAS-trun)

There are seven sea turtle species *(SPEE-sheez)*. The leatherback *(LETH-ur-bak)* sea turtle is the largest *(LAR-jest)*. Its long front flippers can span nearly nine feet, and its black carapace *(KAIR-uh-pase)* is made of rubbery skin reinforced *(ree-in-FORST)* by thousands of small, bony *(BOH-nee)* plates. The smallest species are the olive *(AH-liv)* ridley *(RID-lee)* and the Kemp's ridley, with shells less than 30 inches long.

Dolphins are marine *(muh-REEN)* mammals. These are animals that live in the water, breathe air, give birth to fully formed young *(yung)*, and produce *(proh-DOOS)* milk to feed them.

Like all mammals, dolphins are warm-blooded. This means that their bodies maintain a constant temperature *(TEM-pur-uh-chur)* that is usually *(YOO-zhoo-uh-lee)* warmer than their surroundings.

> **CLOSE-UP**
>
> **Appendages** *(uh-PEND-uh-jez)*
>
> The side pectoral *(pek-TOR-ul)* flippers help a dolphin slow down and steer. The dorsal *(DOR-sul)* fin, on its back, is used for balance. The tail, called the fluke *(flook)*, is used to propel *(pruh-PEL)* a dolphin forward through the water.
>
> **DORSAL** *(DOR-sul)* **FIN**
>
> **FLUKE** *(flook)*
>
> **PECTORAL** *(pek-TOR-ul)* **FIN**

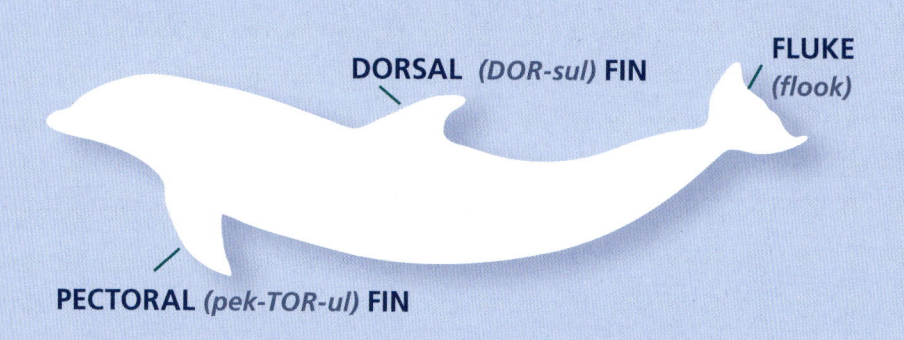

There are about 40 species *(SPEE-sheez)* of dolphin, though this number continues *(kun-TIN-yooz)* to change as new discoveries are made. Dolphins can be found in all the world's oceans *(OH-shenz)* and in some Asian *(AY-zhun)* and South American rivers.

The largest member of the dolphin family is the orca *(OR-kuh)*, or killer whale—which is not a whale at all!

CHAPTER TWO

WELCOME TO THE WORLD

On the sandy shore of Sikopo *(sih-KOPE-oh)*, night has fallen, and a full moon hangs in the sky. A mound of dry sand stirs. Dozens of tiny brown hatchlings emerge from beneath the sand. Their vision is weak, but they see the round, white moon reflecting on the ocean's surface. Running across the sand with their little flippers, the hatchlings race toward the moonlight.

Female *(FEE-male)* turtles go ashore at night to dig holes in the sand and lay a clutch, or group *(groop)*, of about 65 to 100 eggs. Then the turtle covers *(KUH-vurz)* the eggs with sand and returns to the sea. The baby sea turtles, called hatchlings, are left on their own. They will never know their parents.

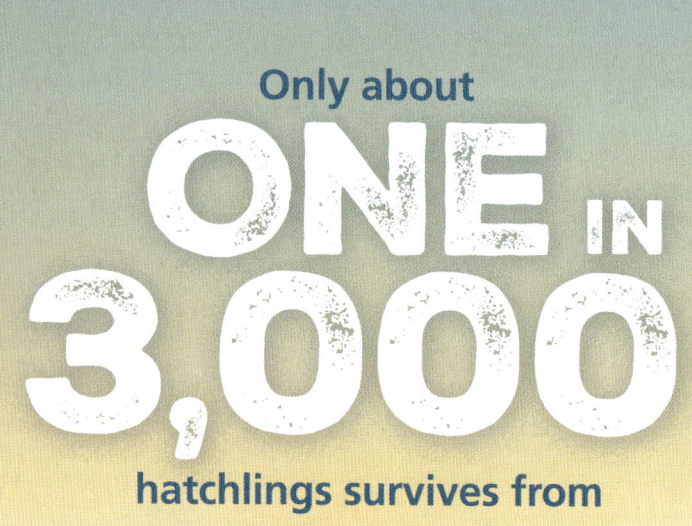

Only about **ONE IN 3,000** hatchlings survives from the nest to adulthood.

The life of a recently hatched sea turtle is filled with danger *(DANE-jur)*. On the shore, seabirds and crabs race to grab the hatchlings. Around reefs and in shallow water, sharks and other large fish may devour *(dee-VOW-ur)* the young turtles. The safest place for hatchlings is far out at sea where the ocean is mostly devoid of large predators *(PREH-duh-turz)*.

The beaches where sea turtles hatch and the currents they ride vary by species. Some young hatchlings swimming at sea are picked up by the South Equatorial *(eh-kwuh-TOR-ee-ul)* Current, which carries them hundreds of miles out to sea. There they may encounter a dense carpet of seaweed.

These floating mats of seaweed provide the perfect habitat for a growing sea turtle. Safely hiding under the canopy *(KAN-uh-pee)* of seaweed, a turtle will poke its nose up for a quick breath *(breth)* every few minutes. Spotting a tiny *(TY-nee)* shrimp clinging to the seaweed, the young hatchling will snatch the creature *(KREE-chur)* with its beak and swallow its first meal.

In the warm ocean near Florida *(FLOR-rih-duh)*, a mother Atlantic spotted dolphin is swimming in broad *(brawd)* circles, flexing her body. The members of her pod surround her, protecting her from any predators. A small tail emerges *(ee-MUR-jez)* from the mother's body. In a few minutes, the rest of her baby slips into the water. Momentarily *(moh-men-TAIR-ih-lee)* unsure what to do, the wiggling calf *(kaf)* begins to sink. His mother immediately *(im-MEE-dee-et-lee)* pushes the newborn toward the water's surface to take its first breath *(breth)* of air.

Dolphins live in groups *(groops)* called pods. Nursery pods consist of only related females and their young offspring. Because calves *(kavz)* are not strong swimmers, they ride pressure *(PRESH-ur)* waves created *(kree-AY-ted)* by their mothers' swimming, like underwater surfing!

Members of nursery pods work together to protect youngsters *(YUNG-sturz)*. However, roughly *(RUF-lee)* 20 percent of calves do not survive their first year. Many are sickened by parasites *(PAIR-uh-sites)* or pollution *(puh-LOO-shun)* or attacked by predators.

HOW LONG CAN A DOLPHIN HOLD ITS BREATH?

1-2 MINUTES
....
BABY DOLPHIN

10-15 MINUTES
....
ADULT DOLPHIN

A baby dolphin can hold its breath for one to two minutes. Adults can stay underwater for ten to fifteen minutes. Dolphins often rest motionless *(MOH-shun-les)* at the surface. They alternate which half of the brain is sleeping so they never lose *(looz)* consciousness *(KON-shus-nes)*.

To strengthen their breathing muscles *(MUS-ulz)*, calves may chin-slap: They raise their heads straight out of the water, take a big breath, and then slap their chins down, forcing air out of their blowholes. Mothers monitor their calves as they chin-slap, nudging *(NUJ-ing)* them to the surface when they slip too far below the waves.

CLOSE-UP

Blowhole

A dolphin breathes through its blowhole, a type of nostril located on top of the head. At the surface, the dolphin opens its blowhole to quickly exhale and inhale fresh air. Then the blowhole automatically shuts.

CHAPTER THREE

MEALTIME

Sea turtles have varied *(VARE-reed)* diets *(DY-ets)* based on their different habitats. Leatherbacks feed mostly on jellyfish that drift on ocean currents. Kemp's and olive ridleys hunt invertebrates *(in-VUR-tuh-brayts)*, such as jellyfish and sea worms, in shallow coastal waters. Hawksbills feed primarily on sponges *(SPUN-jez)* found at coral reefs. Green sea turtles have a completely vegetarian *(vej-uh-TAIR-ee-un)* diet of algae *(AL-jee)*, seaweed, and other plants. Loggerheads forage *(FOR-ij)* for clams, mussels, and crabs on the seabed.

CLOSE-UP

Cleaning Station

Animals form partnerships at coral *(KOR-ul)* reef cleaning stations. "Cleaner fish" and shrimp pick dead skin and harmful animals or plants off larger animals' bodies. This helps the animals stay healthy, and the cleaners get a free meal.

Sea turtles have a highly developed sense of smell because of a special area *(AIR-ree-uh)* on the roof of the mouth called the Jacobson's organ. This organ helps the turtles find tasty food to eat by detecting the scent of small fish, crustaceans *(krus-TAY-shunz)*, jellyfish, worms, and algae.

CLOSE-UP

Beak

Sea turtles do not have teeth. To eat, they use their hard, sharp beak to slice off mouthfuls of food or to crush the shells of prey *(pray)*. Beaks are made of keratin *(KAIR-uh-tin)*—the same substance found in human *(HYOO-men)* fingernails.

Dolphin calves *(kavz)* drink only their mother's milk until they are between three and six months old. By the time they are nine months old, they no longer drink milk and can hunt for fish instead *(in-STED)*.

One hunting technique *(tek-NEEK)* a mother teaches her calf is crater *(KRAY-tur)* feeding. Using echolocation *(eh-koh-loh-KAY-shun)*, a calf can locate a pearly *(PUR-lee)* razorfish *(RAY-zur-fish)* hiding under the sand. Positioning its body vertically, it begins to spin. Its rostrum *(RAW-strum)*, or snout, digs a crater into the sand. The calf then stuns the uncovered fish with a loud burst of sound and snaps up the fish.

CLOSE-UP

Echolocation *(eh-koh-loh-KAY-shun)*

Dolphins send out high-pitched sound waves. These waves bounce off an object and return to the dolphin as an echo *(EH-koh)*. This echo creates a mental picture of the size and shape of the object. Like an X-ray, echolocation can also be used to see inside things.

CLOSE-UP
Teeth
Dolphins' teeth are cone-shaped and of uniform *(YOO-nih-form)* size. The teeth interlock, securely *(seh-KYUR-lee)* gripping prey. Dolphin teeth are permanent. Lost or broken teeth are not regrown *(ree-GRONE)*.

Dolphins feed mostly on fish, squid *(skwid)*, and shrimp. Some species have as many as 250 teeth, yet dolphins don't chew their food. They use them to trap prey, and then swallow it whole *(hole)*.

Dolphins living in different habitats possess *(puh-ZES)* different skills. Teams of dusky dolphins use sound to corral *(kor-RAL)* anchovies *(AN-choh-veez)* into a tight ball. Then they rush through the ball and grab mouthfuls of the fish.

CHAPTER FOUR

STAYING SAFE

As they roam the world's oceans, sea turtles follow currents. These fast-flowing streams help sea turtles travel widely without expending a lot of energy. Without currents pushing them along, leatherback sea turtles can swim about six miles per hour, but other sea turtles max out at less than two miles per hour. Sea turtles cannot outpace predators in the water. They have a myriad *(MEER-ee-ad)* of enemies, including large fish, from sharks and groupers *(GROOP-urz)* to mackerels and barracudas *(bare-uh-KOO-duz)*.

Adult sea turtles are better protected by their shells than young turtles. But they are still no match for orcas *(OR-kuz)* and large sharks. Leatherbacks brave the open ocean. Most other sea turtles live along coasts or around coral reefs, where large sharks are less common. Ledges and rock crevices *(KREV-ih-sez)* also provide hiding places for sea turtles in these areas. If sea turtles can survive the many perils *(PAIR-ulz)* of life at sea, they can live to be 50, 70, or even 100 years old.

Dolphins have few enemies in the sea but will sometimes face off with hungry sharks. And even though orcas are dolphins themselves, orcas do attack and eat smaller dolphins when necessary *(NEH-sis-sair-ee)*. Communication *(kuh-myoo-nih-KAY-shun)* is vital *(VY-tul)* to dolphins' survival against these enemies. Using a language *(LAN-gwij)* that includes whistles *(WIH-sulz)*, clicks, and other sounds, they can cooperate *(koh-OP-ur-ate)* to drive them away.

Communication between dolphins is used in other ways as well. It helps dolphins form relationships and hunt cooperatively *(koh-OP-ur-uh-tiv-lee)*. Each dolphin has a signature *(SIG-nuh-chur)* whistle that it uses like a name to identify itself.

The Wild Dolphin Project is an organization *(or-geh-nih-ZAY-shun)* that has developed something like a dictionary *(DIK-shun-air-ee)* of dolphin sounds. It is working on a computer-aided *(kum-PYOO-tur-AY-ded)* device that may allow humans to "speak" to dolphins.

CLOSE-UP

Communication

Dolphins communicate and form social *(SOH-shul)* bonds by rubbing and nudging each other. They also whistle through their blowhole and click using a nasal *(NAY-zul)* sac in their forehead, called a melon. They can make squeaks *(skweeks)*, quacks, grunts, barks, and trills. Most dolphin sounds are too high-pitched to be heard by human ears.

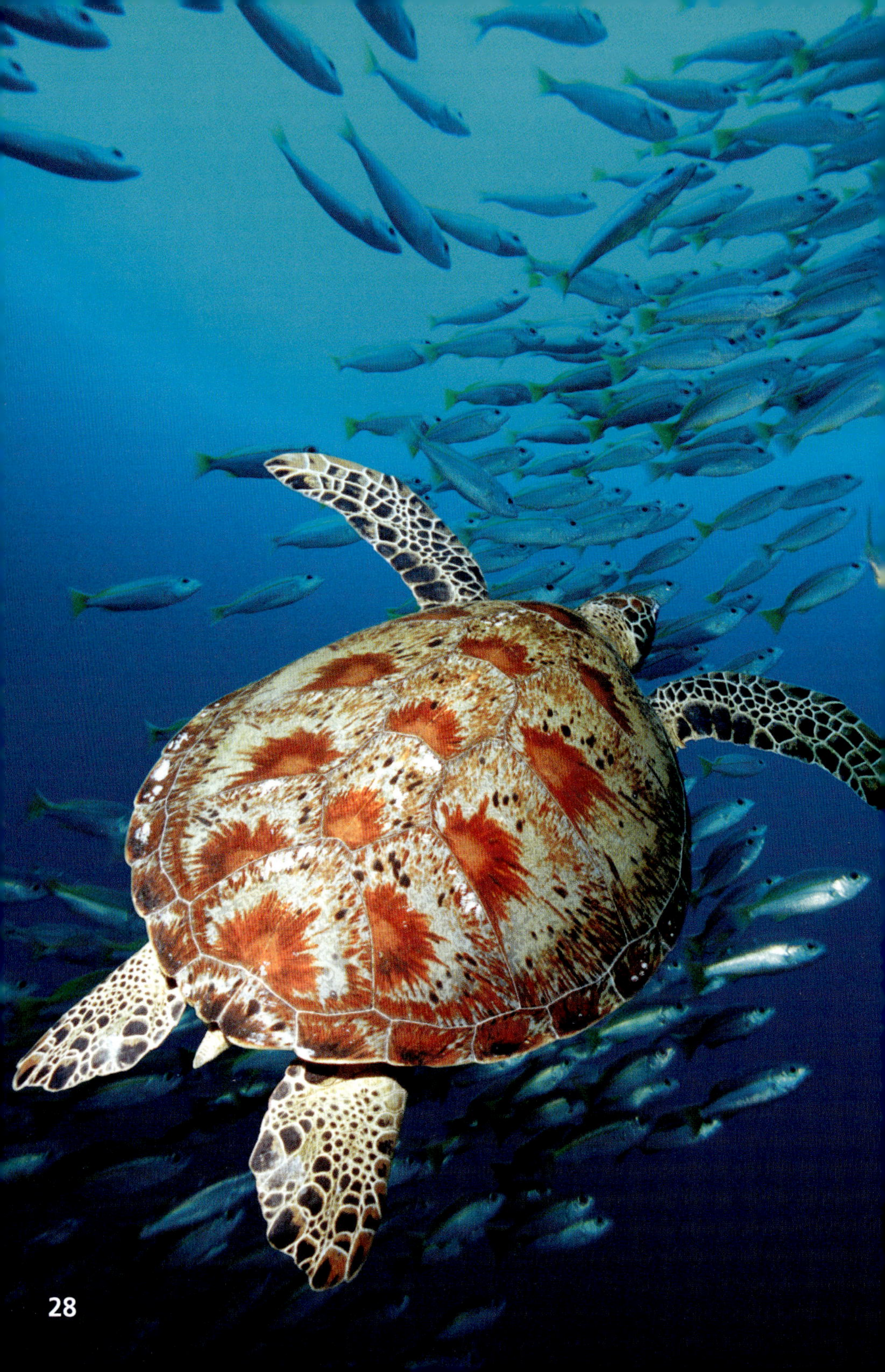

Humans can sometimes be the enemy as well. All seven species of sea turtle are endangered *(en-DAYN-jurd)*. Rising water temperatures and pollution *(puh-LOO-shun)* are making them sick, and they sometimes mistake garbage *(GAR-bej)* for food.

Many sea turtles drown when they get caught *(kawt)* by huge fishing nets. As humans continue *(kun-TIN-yoo)* to expand their use of the oceans, sea turtles will need more of our help to survive.

Some dolphin species are also in trouble *(TRUH-bul)*. River and coastal dolphins suffer habitat destruction. Chemicals *(KEM-ih-kulz)*, sewage *(SOO-wij)*, oil, and other poisons pollute *(puh-LOOT)* their homes.

Ganges *(GAN-jeez)* River dolphins live in one of the most polluted rivers on Earth. Now, fewer than 2,000 exist in their native India *(IN-dee-uh)*.

Oceanic *(oh-shee-AN-ik)* dolphins face challenges as well. They often get caught in giant nets used for large-scale fishing. Learning what dolphins need to thrive is one way to help protect them.

Dolphins are believed *(bee-LEEVD)* to be **SMARTER** than chimpanzees!

CHAPTER FIVE

COOL FACTS

CLOSE-UP

Shedding Skin

Dolphins constantly shed dead skin. Some dolphins replace a thin outer layer of skin every two hours! This keeps their skin smooth and parasite-free. Their skin is also wavy, which reduces *(ree-DOO-sez)* drag, so dolphins can travel up to 25 miles per hour in short bursts.

> **CLOSE-UP**
>
> **Salty Tears**
>
> Sea turtles' saltwater habitats create a buildup *(BILD-up)* of excess salt in their bodies. Their bodies remove salt by seeping salty fluid *(FLOO-id)*, similar to tears, out of tiny holes near their eyes.

For young dolphins, living in a group, or pod, is like going to school. They learn cooperative *(koh-OP-ur-uh-tiv)* hunting strategies *(STRAT-uh-jeez)* and develop social *(SOH-shul)* skills. And they play with each other, which engages *(in-GAY-jez)* their natural *(NACH-ur-rul)* curiosity *(kyur-ee-AH-sih-tee)* and stimulates their brains. One "game" involves a group of dolphins tossing seaweed into the air, dragging it around in their mouths or on their flippers, and passing it to other dolphins.

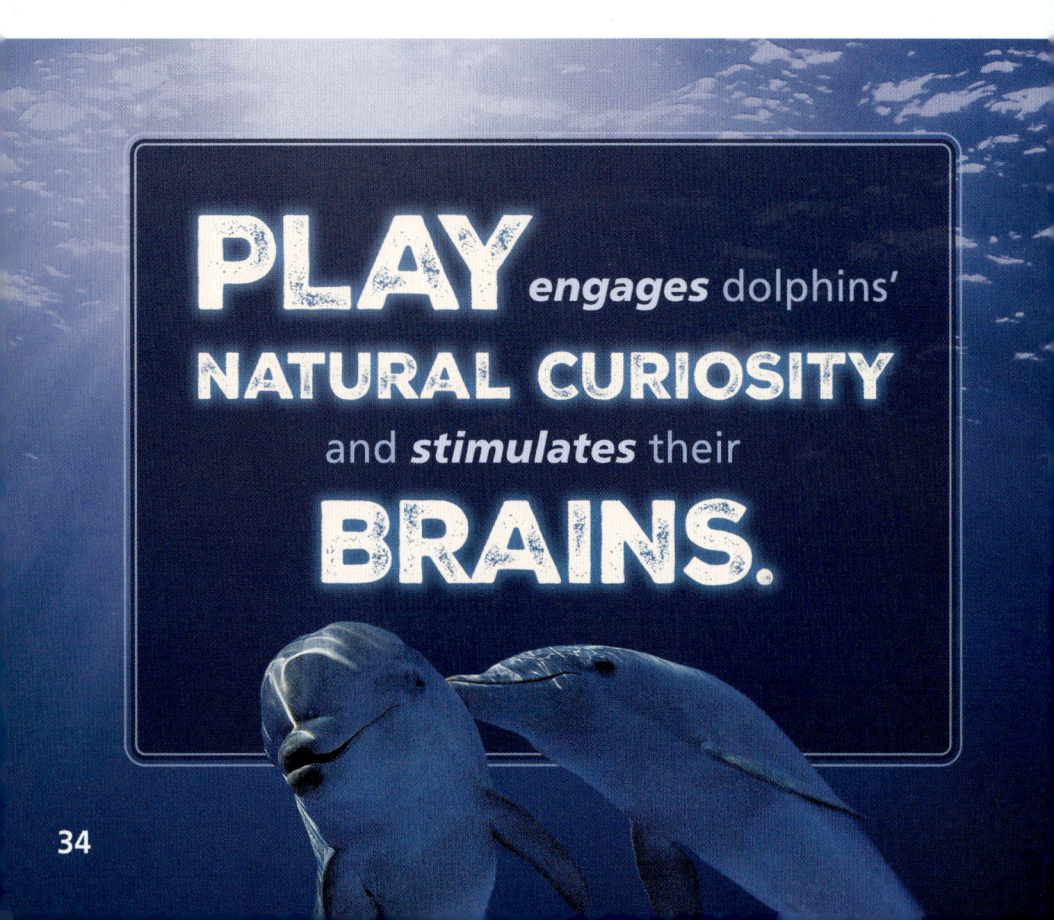

PLAY *engages* dolphins' NATURAL CURIOSITY and *stimulates* their BRAINS.

At one year old, a sea turtle can stay underwater for several minutes. An adult turtle can stay underwater for up to five minutes. A resting or sleeping sea turtle can stay under for six hours! Spongy *(SPUN-jee)* tissues *(TISH-ooz)* in their mouth and throat absorb oxygen *(OX-ih-jen)* from the water.

Sea turtles hold their BREATH for 4 TO 5 MINUTES.

Dolphins and sea turtles are just two of the hundreds of thousands of animals that make the oceans, lakes, and rivers of the world their home. Continued research *(REE-surch)* and study will help preserve and protect these beautiful animals and habitats for generations *(jen-ur-AY-shunz)* to come.

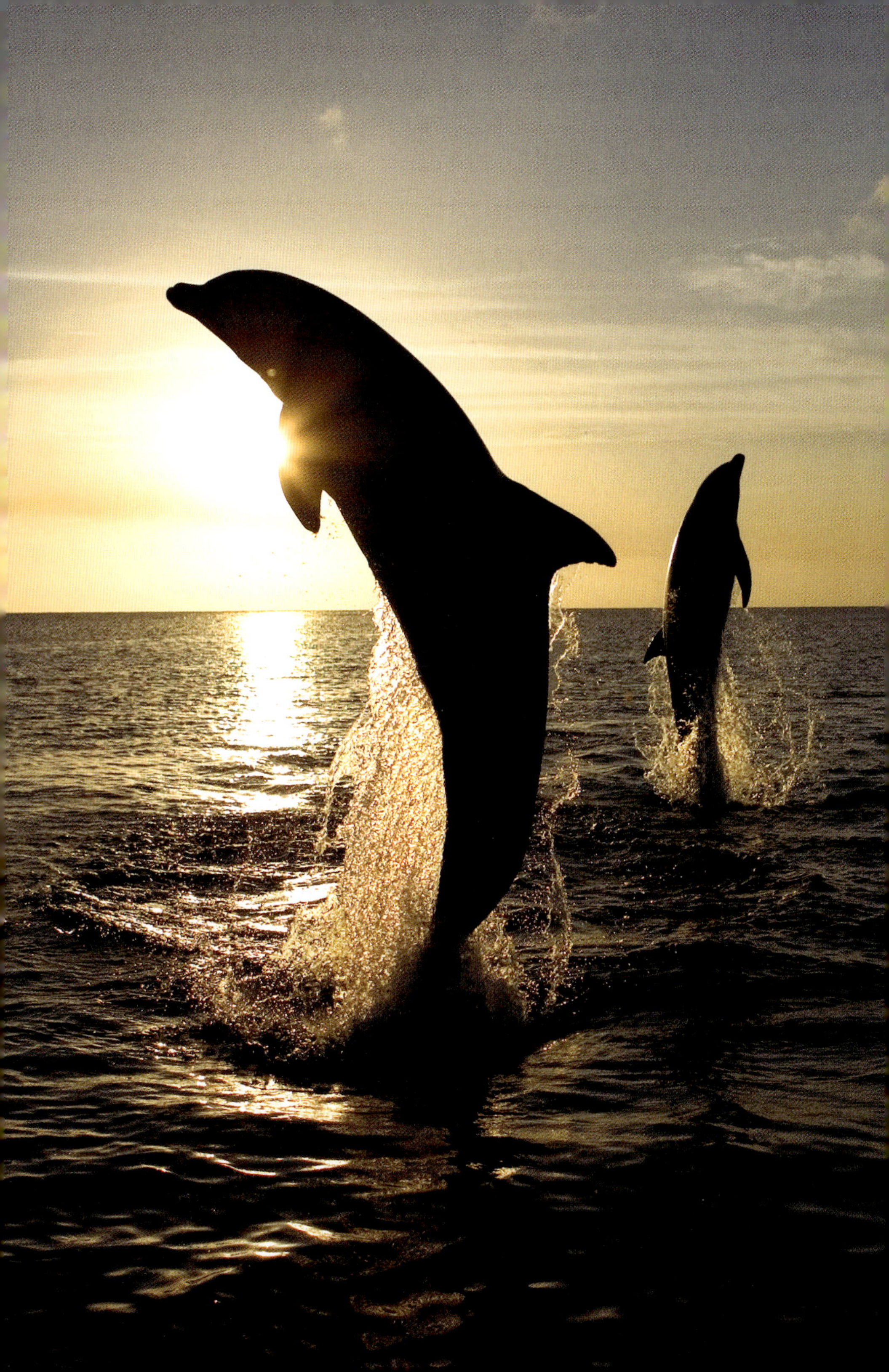

CHAPTER SIX

FAMILY ALBUM SNAPSHOTS

Here a just a few of the different species of sea turtles and dolphins:

Green sea turtles are widespread in warm subtropical and tropical ocean waters. They nest in more than 80 countries *(KUN-treez)* and are about the same size as loggerheads.

Leatherback sea turtles are more tolerant of cold water than other sea turtles. They have been spotted as far north as Canada *(KAN-uh-duh)* and Norway and as far south as New Zealand.

Hawksbill sea turtles mainly inhabit warm coral *(KOR-ul)* reefs in the Atlantic, Indian *(IN-dee-in)*, and Pacific oceans. Their shells are about 35 inches long, and they weigh *(way)* about 150 pounds.

Olive ridley sea turtles primarily nest on warm Pacific shores of the Americas and along the northeastern coast of South America. They prefer warmer waters than Kemp's ridleys.

Loggerhead sea turtles weigh up to 450 pounds, with shells about 4 feet long. They can be found in the Mediterranean *(meh-dih-tuh-RAIN-ee-en)* Sea and the Atlantic, Pacific, and Indian oceans

The endangered *(en-DAYN-jurd)* **Irrawaddy** *(EER-uh-wod-dee)* **dolphin** lives near the coast and estuaries *(ES-choo-air-eez)* of the Bay of Bengal *(ben-GAHL)* and Southeast Asia.

Commerson's *(KOM-ur-sunz)* **dolphin** of coastal southern Argentina *(ar-jen-TEEN-uh)* is nicknamed the "panda dolphin" or "skunk dolphin" for its striking black and white coloration *(kuh-lur-RAY-shun)*.

Spinner dolphins have 250 teeth, the most of any mammal. They are also the only dolphin to naturally *(NACH-ur-uh-lee)* leap from the water in corkscrews *(KORK-skrooz)*.

The **false killer whales** live in Hawaiian *(huh-WY-en)* waters. In 2012, they were listed as an endangered species.

The **Amazon River dolphin**, also called boto *(BOH-toh)*, has a flexible neck to help it maneuver *(muh-NOO-vur)* through tangles of underwater plants and tree roots.

GLOSSARY

carapace *(KAIR-uh-pase)*: the hard upper shell of a turtle or crustacean

clutch: a group of eggs produced and incubated at the same time

crustaceans *(krus-TAY-shunz)*: animals with no backbone that have a shell covering a soft body

endangered *(en-DAYN-jurd)*: at risk of disappearing from Earth forever

invertebrates *(in-VUR-tuh-brayts)*: animals that lack a backbone, such as shellfish, insects, and worms

parasites *(PAIR-uh-sites)*: animals or plants that live on or inside another living thing (called a host) while giving nothing back that the host needs

rostrum *(RAW-strum)*: a stiff, beaklike snout extending from an animal's head

species *(SPEE-sheez)*: a group of living beings with very similar features

estuaries *(ES-choo-air-eez)*: coastal bodies of water where fresh water from rivers and streams mixes with salt water from the ocean

QUESTIONS

1. There are several marine *(muh-REEN)* mammals other than dolphins. Can you think of any other mammals that spend much of their time in the ocean?

2. Sea turtles are one of the few *reptiles* that live in the ocean. Can you think of any reptiles that spend much of their time in rivers or lakes? How might you learn more about these reptiles?

3. Reptiles are cold-blooded animals. What are some ways that cold-blooded animals regulate *(REG-yoo-late)* their body temperature, so they don't get too hot or too cold?

4. Researchers are looking for ways to communicate *(kum-MYOO-nih-kate)* with dolphins. Do you think dolphins use language *(LANG-gwij)* the way humans do? Do you think humans and dolphins will someday be able to "speak" to each other?

5. What are some of the dangers young turtles and dolphins face while growing?

CODE READER™
Making Difficult Words Easy

Code Reader Books provide codes with "sound keys" to help read difficult words. For example, a word that may be challenging to read is "chameleon," so it might be followed by a code like this: chameleon *(kuh-MEE-lee-un)*.

The codes use phonetic keys for each sound in the word. Knowing the keys can help make reading the codes easier.

Code Reader™ Keys

Long a sound (as in make):
a *(with a silent e)*, **ai**, or **ay**
Examples: break *(brake)*; area *(AIR-ee-uh)*; able *(AY-bul)*

Short a sound (as in cat): **a**
Example: practice *(PRAK-tis)*

Long e sound (as in keep): **ee**
Example: complete *(kum-PLEET)*

Short e sound (as in set): **e** or **eh**
Examples: metric *(MEH-trik)*; bread *(bred)*

Long i sound (as in by):
i *(with a silent e)* or **y**
Examples: might *(mite)*; bicycle *(BY-sih-kul)*

Short i sound (as in sit): **i** or **ih**
Examples: myth *(mith)*; condition *(kun-DIH-shun)*

Long u sound (as in cube): **yoo**
Example: unicorn *(YOO-nih-korn)*

Short u or schwa sound (as in cup):
u or **uh**
Examples: pension *(PEN-shun)*; about *(uh-BOWT)*

Long o sound (as in hope):
o *(with a silent e)*, **oh**, or **o** at the end of a syllable
Examples: molten *(MOLE-ten)*; ocean *(OH-shen)*; nobody *(NO-bah-dee)*

Short o sound (as in top): **o** or **ah**
Examples: posture *(POS-chur)*; bother *(BAH-ther)*

Long oo sound (as in cool): **oo**
Example: school *(skool)*

Short oo sound (as in look): **ŏŏ**
Examples: wood *(wŏŏd)*; could *(kŏŏd)*

oy sound (as in boy): **oy**
Example: boisterous *(BOY-stur-us)*

ow sound (as in cow): **ow**
Example: discount *(DIS-kownt)*

aw sound (as in paw): **aw**
Example: faucet *(FAW-sit)*

qu sound (as in quit): **kw**
Example: question *(KWES-chun)*

zh sound (as in garage): **zh**
Example: fission *(FIH-zhun)*